# HOW TO BECOME WEALTHY BEFORE 30

## *The Perfect Guide to Become Rich Before 30*

Peter McKenna

TABLE OF CONTENTS

# INTRODUCTION

Being a billionaire before the age of 30 might seem like an unachievable dream. That is no longer accurate, though. Making wise financial decisions early in your life positions you to eventually become a millionaire. Saving money today, however, does not always equate to becoming a millionaire. You must devise a variety of strategies to assist you in achieving the goal. Being a billionaire involves more than just having a large number of zeros in your bank account. Although some people might not be comfortable with investing money, this is not a barrier to becoming a billionaire. To progress from a life of squalor or nothing at all to one of luxury is the typical American ideal. If you want to be a billionaire, take chances, make wise investments, and cling onto your money.

## Chapter 1: The Right Money Mindset
### 1.1 What is Money Mindset?

A money mindset is an overarching perspective you hold on money. It influences the important financial decisions you make every day. And it can significantly affect your capacity to carry out your objectives. If you alter your perspective on money, you'll probably choose more effective solutions to problems. The power of positive thought is crucial in this situation. Your financial mindset acts as an impetus for action, similar to emotional inertia. You are more likely to be resolute and take the actions necessary to achieve if you have a good attitude toward money. Conversely, negativity fosters feelings that keep people from taking action, such as fear or intimidation, defeatism, and procrastination.

The difficulties seem too great to overcome, so it seems simpler to just ignore your financial situation. When viewed in this way, your perception that you are incapable of improving yourself is what is preventing you from succeeding, not debt, a poor credit score, or a meager income.

Past financial experiences have a big impact on how we think about money today. It's completely possible that your parents or grandparents taught you some basic financial principles, but it's also possible that you grew up in a home where talking about money was frowned upon. From there, based on your life experiences, you've either consciously or unconsciously created a specific set of views about money. Regardless of your upbringing, the information you receive today can either enhance or change your financial perspective. By adopting a more optimistic view and regaining control over the aspects of your financial life that you have the capacity to influence, you can most surely develop new attitudes and reform old ones. A negative financial mindset often produces even worse financial mindsets. It's quite simple to catastrophize if you get caught up in a downward thought spiral. You might

think you'll never be "good at money" or
experience anxiety at the thought of putting
in the effort necessary to money" or
experience anxiety at the thought of putting
in the effort necessary to improve.

Your financial mentality is the problem.
Your financial mindset affects how much
you can and should earn, how you should
use debt, how much you can give away, and
how confidently and successfully you can
invest. It also affects what you believe you
can and cannot do with money.
Another sign of your money mindset is how
you behave in financial situations: do you
feel weak and intimidated or confident and
in control? Are you confident in asking
questions? If so, then you probably feel safe
doing so. If not, then you probably feel shy
and embarrassed. Do you unconsciously
believe that money belongs to males and
that women shouldn't worry about it?
Our unconscious mind is where our
fundamental notions about money exist,

which is fascinating. It may also remain there, inactive. However, by learning to pay close attention to your thoughts, feelings, physical responses, and interactions with money, you can become more aware of your present set point and develop the capacity to alter it.

## 1.2 Reset your mindset about money

Observing and internalizing the financial lessons we gained from our parents, friends, community, and other caregivers helped shape many of our core attitudes about money during our formative years. in particular, our parents. Young kids are like tiny sponges. We paid close attention to how the adults closest to us handled money and other aspects of life. Did your parents fight over money, or were they cool under pressure when the topic of family finances came up?

You may still harbor an ingrained belief that "people like us don't succeed" if you were

raised in a household that held such view.
You might feel extremely uncomfortable if
you make more money than your parents do.

Your financial perspective is positively
impacted by growing up in material comfort
with thoughtful, giving parents who
communicate with one another and other
people in a courteous and peaceful manner.
It's easier to change your money mindset if
you know where it came from. You want to
know and understand your current beliefs
about money, just like you would before
setting out on a journey and entering your
coordinates into the navigator.
Knowing where you started will help you
appreciate how far you've gone as you
change your mentality. It is like knowing
you have driven 200 miles when you glance
at the navigator. Money is different, though,
in that it involves a lifetime's worth of
travel. We never actually get to where we're
going. We simply keep expanding.

Being aware is the first step in changing your money perspective. Pay close attention to how you think, act, and behave when it comes to money. Your moods will be influenced by how you think about money. Your action is influenced by your sentiments. So, for instance, if you believe that money is a scarce resource, you'll experience tension and worry. You can become angry with members of your family for their monetary frivolity, negligence, or errors. You won't be kind, and you might even accumulate wealth and material items in a hoard.

Keep in mind that your attitude about money has nothing to do with how much money you make or actually have. The belief that there is not enough money to go around and that they will never have enough is shared by some wealthy people. There are also others with little money and goods who have a deep-seated sense of blessing, abundance, and richness. You'll come across your own money blockages and limiting ideas that prevent change as you become more conscious of your mentality. Because of these ideas, you are unable to feel and behave more abundantly. This is entirely typical. Everybody has financial blocks. They remain forever. Money obstacles and limiting ideas constantly coming back in fresh and varied forms. It is your responsibility to consistently identify the barriers, remove them, and let go of them in order to develop a healthier and more secure relationship with your money.

How to change your financial perspective;

**1. Confirm your control over the situation.**

Everybody has a different upbringing and a unique set of circumstances. Keeping that in mind, you are in charge of the daily financial decisions you make. At any level of wealth building, having the sense that you are in charge of your finances and have the power to improve them may be quite beneficial.

## 2. Be prepared to work hard.

There are tons of free tools available online if you are unclear about specific personal financial subjects. You don't have to fully comprehend every subject, but there are educational tools available so that you can grasp the fundamentals. You will place yourself at a distinct edge if you are just willing to learn.

## 3. Appreciate small victories.

Recognize your accomplishments when you pay off a little debt or contribute to a retirement plan offered by your work. Over time, modest improvements might result in huge advancements. Small victories are another sign that your actions are consistent with your larger objectives.

## 4. Make a commitment to success.

There are good days, months, and years for everyone. In life, a lot of things do occur by chance and are ultimately beyond your control. In these situations, maintain an optimistic attitude and reiterate your dedication to achieving your financial objectives despite setbacks.

## 5. Show your appreciation.

Nobody's financial situation is ideal, and it is safe to assume that almost everyone experiences some degree of financial anxiety. Be appreciative for what you have already accomplished as well as the fact that you are making every effort to enhance your financial situation.

## 6. Stop self-critical thoughts before they spiral.

When you start telling yourself you cannot accomplish anything financially, one strategy to use is to figuratively picture a stop sign. The truth is that you can accomplish your goals, but having self-confidence can be quite difficult if you have limiting beliefs. Make every effort not to criticize yourself.

## 7. Continue on your path.

We are who we are because of our experiences, but we must have the courage to move forward despite the past. Investing with a long-term perspective can help you develop good habits now. Putting money into long-term investments like mutual funds or ETFs can give your financial life direction and a certain amount of tranquility.

## 8. Develop a mindset of abundance.

It is not your loss if someone else succeeds. Remind yourself that you are a completely distinct person with a completely different set of life circumstances and experiences, even if you feel like others are "ahead" of you in the money game. A mindset of abundance, as opposed to one of scarcity, might serve as a reminder that there is plenty space for everyone to shine.

# Chapter 2: Set Financial Targets

## 2.1 Why the Rich Set Financial Targets

Setting financial goals is essential for accumulating wealth. Studies have indicated that goal-setters and "planners" are more financially successful and feel better about their financial status than those who don't make any preparations, regardless of income level. Financial objectives resemble the itinerary receipt for a flight. They outline your desired destination, the cost of the "travel," and the anticipated arrival time.

Financial objectives can help you focus on and reduce down your options while making financial selections. Setting short-term, mid-term, and long-term financial goals is a crucial first step toward achieving financial security. Without a clear goal in mind, you run the risk of going overboard with your spending. Then, when you need money for unforeseen expenses, not to mention when you want to retire, you won't have enough.

It's possible to become trapped in a cycle of credit card debt and feel as though you'll never have enough money to get adequately insured, making you more exposed than you need to be to tackle some of life's biggest hazards.

As the world learned during the epidemic and as many families discover each month, even the most prudent person cannot be fully prepared for every disaster. By planning ahead, you have the opportunity to consider potential outcomes and make every effort to be ready for them. This needs to be a continuous process so that you can adapt your life and ambitions to the changes that are unavoidably going to occur. Your opportunity to formally examine your goals, change them, and assess your advancement from the previous year is provided by annual financial planning. Consider creating goals if you haven't already so you can establish or maintain a solid financial foundation. To help you learn to live comfortably within your means, lessen your financial problems, and save for retirement, financial experts advise setting the following goals, ranging from short-term to long-term.

The beauty of yearly financial planning is that through the difficulties of life, you can evaluate, update, and track your progress toward your goals. You'll discover that throughout the process, both the little things you do every day and every month and the bigger things you do every year and over many years will help you reach your financial goals.

## 2.2 How to Create Financial Targets

1. Determine what is important to you. Place everything from the urgent and necessary to the frivolous and distant on the table for evaluation and consideration. Determine what can be done right now, what will take some time, and what has to be done as part of a long-term plan.

2. Use the SMART-goal approach. To put it another way, make sure your goals are Timely, Specific, Measurable, Achievable, and Relevant. SMART.

3. Make a sensible spending plan. Gain a firm grasp on what is coming in and what is leaving, then use it to further your goals. To stop leaks in your financial ship, use your budget. I hope that your strict, practical, watertight budget will reveal at least a few dollars that are left. Whatever it is, set up an automated transfer into a different account that will take care of the first few items on your list of priorities.

4. Follow your development. Verify that you are meeting the required standards. If not, give it some thought again and figure out what went wrong.

5. Making a strategy that prioritizes your goals is the best method to achieve your financial objectives. You'll find that some of your own goals are vast

and far-reaching, while others are specific and smaller in scope.

## 2.3 Factors to Consider When Setting Your Targets

### Feasibility

A goal that is too simple to complete can make you feel accomplished before you are ready, while a goal that is too tough to complete can suck up resources unnecessarily. The value of a goal to your business is significantly influenced by its viability. Create attainable goals that will help you expand your firm without going out of business by using historical data, current market trends, and an examination of your company's resources.

### Time Frame

The approaches you take to accomplish a certain company goal will rely on how much time you give yourself to do it. While a goal with a longer time horizon could be more expensive, it might also wind up paying off more in the long run. However, a shorter period can let you test out fresh business concepts while limiting any potential drawbacks.

**Resources**

The resources of your business play a big role in the objectives you establish. You might need to hire extra staff if you don't have enough of them to complete a task. However, a financial analysis may show that you are now unable to afford to hire more workers. Setting and prioritizing goals and building business plans both need careful consideration on how to balance resources.

**Results**

Setting goals requires consideration of the outcomes of earlier business plans. Create a performance-measurement system to aid in assessing the viability of your business plan. To assist create more successful business strategies in the future, learn from the mistakes and keep track of the triumphs.

## 2.4 Tips on How to Achieve Your Financial Target Easily

**Set measurable objectives.**
Your objectives should be something that you can see yourself making progress toward and something that you can complete in a reasonable amount of time. Being a millionaire is a definite and measurable goal, but it might not be one that can be achieved very soon. Large goals should be divided into smaller, more achievable pieces. For instance, setting a SMART goal to raise your net worth by 10% can help you get closer to your ultimate objective of becoming a millionaire by regularly accomplishing that target.

## Match Your Objective to Your Values

Saving money for emergencies and retirement is beneficial for almost everyone, but beyond that, be sure your financial objectives align with your values. If you genuinely don't want to settle down in one place, it could be difficult to save for a down payment on a house, for example. In that scenario, it could be preferable to pay your rent and devote your attention to something that matches your ideal way of life, like setting aside money for travel or pursuing a life as a digital nomad.

## Resources

The resources of your business play a big role in the objectives you establish. You might need to hire extra staff if you don't have enough of them to complete a task. However, a financial analysis may show that you are now unable to afford to hire more workers. Setting and prioritizing goals and building business plans both need careful consideration on how to balance resources.

## Results

Setting goals requires consideration of the outcomes of earlier business plans. Create a performance-measurement system to aid in assessing the viability of your business plan. To assist create more successful business strategies in the future, learn from the mistakes and keep track of the triumphs.

## 2.4 Tips on How to Achieve Your Financial Target Easily

**Set measurable objectives.**

Your objectives should be something that you can see yourself making progress toward and something that you can complete in a reasonable amount of time. Being a millionaire is a definite and measurable goal, but it might not be one that can be achieved very soon. Large goals should be divided into smaller, more achievable pieces. For instance, setting a SMART goal to raise your net worth by 10% can help you get closer to your ultimate objective of becoming a millionaire by regularly accomplishing that target.

## Match Your Objective to Your Values

Saving money for emergencies and retirement is beneficial for almost everyone, but beyond that, be sure your financial objectives align with your values. If you genuinely don't want to settle down in one place, it could be difficult to save for a down payment on a house, for example. In that scenario, it could be preferable to pay your rent and devote your attention to something that matches your ideal way of life, like setting aside money for travel or pursuing a life as a digital nomad.

People should set aside time for self-evaluation in order to choose the proper goals. They should consider how they see their future selves and then make decisions today that will lead them toward that future lifestyle.

## Plan Frequent Check-Ins

Being able to observe progress is encouraging. Additionally, it gives you the chance to think about what isn't working and how you may adjust your approach if you aren't making any progress. Additionally, if your financial and personal circumstances change, regular check-ins give you the chance to decide whether your goals need to change as well.

Set aside a set time to evaluate your progress. Depending on what makes sense for the objective you have selected, it could be weekly, monthly, or quarterly.

**Utilize the Correct Tools**

If you are utilizing the appropriate tools to monitor progress, regular check-ins can be quick and simple. They can also aid in keeping your focus on the goal.

To do this, you might hang sticky notes on your bathroom mirror as a reminder or cross tasks off a list that is relevant to your objective.

Numerous apps can also help you keep motivated and keep an eye on your expenses. Other apps have more specialized features, while some, like Mint, are general-purpose budgeting apps. For instance, Personal Capital can track net worth while SoFi Invest makes investing simpler.

**Put yourself first.**

If you put money aside before you even see it, reaching financial objectives will be simpler. That entails setting up direct payroll contributions into an emergency fund or a savings account, as well as starting automatic transfers to specified accounts. Pay attention to the timing if you use automatic transfers. Making ensuring you have money on hand for important expenses like rent or a mortgage is important.

**Find a Partner Who Can Be Accountable**

Even when motivation wanes, having a companion to hold you accountable and provide moral support can help you keep to your goals. Pairing up with someone who shares your objectives may be ideal, but simply finding a friend who will solicit regular updates on your progress can be beneficial.

If you do not know someone in person, look online or look for local meet-up groups.

## Get Rid of Temptation

Make it tough for someone to undermine your efforts no matter what your purpose is. Make it difficult to spend more money, for instance, if you are trying to pay off debt or stay inside a strict budget. Avoid shopping, unsubscribe from mailing lists, and recycle ads without reading them.

Of course, FOMO, or the fear of missing out, may be a significant obstacle to achieving financial objectives. Informing your pals of your goals will prevent them from luring you with invites to pricey outings. Additionally, get used to saying no so that it will come naturally if and when they ask.

**Find a Partner Who Can Be Accountable**

Even when motivation wanes, having a companion to hold you accountable and provide moral support can help you keep to your goals. Pairing up with someone who shares your objectives may be ideal, but simply finding a friend who will solicit regular updates on your progress can be beneficial.

If you do not know someone in person, look online or look for local meet-up groups.

**Get Rid of Temptation**

Make it tough for someone to undermine your efforts no matter what your purpose is. Make it difficult to spend more money, for instance, if you are trying to pay off debt or stay inside a strict budget. Avoid shopping, unsubscribe from mailing lists, and recycle ads without reading them.

Of course, FOMO, or the fear of missing out, may be a significant obstacle to achieving financial objectives. Informing your pals of your goals will prevent them from luring you with invites to pricey outings. Additionally, get used to saying no so that it will come naturally if and when they ask.

## Chapter 3: Healthy Financial Habits to Start Now

It might take a lot of time to stick to your goals and develop better habits, so the sooner you start, the better. Your financial well-being may be equally crucial to your physical wellbeing. Here are five financial habits you may practice today to improve your financial situation in the future:

**Monitor your spending**.

In order to hold oneself accountable, it's critical to have a visual picture of your spending patterns and financial situation. Although it may seem like a laborious chore, it need not be. You can get a notepad and put pen to paper for a more conventional strategy. However, if you find it overwhelming to think of recording every purchase you make, you can think about utilizing a personal financial application, like Money Manager, to keep track of your spending and have full access to your bank account (or numerous bank accounts). You may set priorities and modify your spending habits for a healthier overall budget by keeping track of your expenditures.

**Begin saving little by little.**

Saving money may seem daunting, but doing so is essential if you want to ensure that your future will be financially stable. Begin progressively. Try making a goal in January to save $50 every month for your Christmas fund rather than setting aside $500 of your upcoming paycheck in anticipation of the holiday.

## Consider yourself an investor

Investing is the key to securing a prosperous future. Start investing as soon as you can for the best results. That merely means you'll have more money in your pocket in the long run. If you don't know how to invest, starting with a 401k account will help you start saving for retirement, which can be quite beneficial. Never forget that having extra cash on hand is the ideal situation for investing. Set investing as a goal for the near future if you're currently a little constrained for cash.

## Reselling and store sales

You may want to browse the sale racks if you do not already shop for deals. You don't have to give up flair to stay on a tight budget. You may prepare your closet for succeeding seasons if you shop at the correct retailers and end-of-season sales. Focus on basics rather than trends that can become dated in a few seasons.

**Improve and establish your credit**

To be able to make those big purchases down the road, it is crucial to build your credit score and understand what factors affect it. When determining your mortgage rate, auto payment, and other costs, one of the key variables is your credit score. So that you can safely apply for a loan in the future, begin establishing credit right away and take the essential actions to raise your score.

# Chapter 4: Proper Investment
## 4.1 Tips on How to Start an Investment

**Boost and establish your credit**

Building your credit score and comprehending its influences are essential if you want to be able to make those major purchases in the future. Your credit score is one of the most important factors in deciding your mortgage rate, auto payment, and other charges. Start building credit as soon as possible and take the necessary steps to improve your score so that you can apply for a loan in the future without fear.

**Early Investment**

It is ideal to start investing early. One reason is that you would need less money annually to reach your investment goals the earlier you start. Don't be afraid to start investing, even if you're a college student or, better yet, in your final year of high school. Your earnings will compound over time.

**Automate your investment process**

Set aside a specific sum of money to invest automatically each month. Through a variety of brokerage providers and automated investment programs like Wealth front, you can create automatic investment plans. You can prevent stagnating and consistently make investments by doing this.

**Check your finances**

You must consider your available funds before you can start investing. Be sensible about it. Be sure to give yourself enough money to cover your usual monthly expenses, such as loan payments. There are risks associated with investing, but you don't need a lot of cash to get started. You do not want to be unable to pay other significant bills.

## Understand investing

It is time to begin studying about investing once you have your money in order. Learn the fundamental terms so you can make informed selections. Discover more about mutual funds, CDs, stocks, and bonds. Don't overlook other factors like portfolio optimization, market effectiveness, and diversity.

## Beware of Commissions

The goal of professionals is to get you to invest in goods that will give them a large commission. Don't move forward without doing extensive research. It is well known that some so-called experts sell products that bring them high commissions but provide little value to their clients.

**Make several different investments**

Prices are constantly rising and falling because of the market's constant change. To prevent suffering a substantial loss of wealth when markets collapse, make sure your portfolio is diversified. You will have some stocks rising while others are falling if you do this. Another option is to invest in overseas markets, which are very different from American ones.

**Checkout Your Portfolio**

You should constantly review your portfolio. What is appropriate for your portfolio today might not be so tomorrow. Knowing what you have and where modifications may be necessary in the future is crucial. Be ready to adjust your investments as the economic environment changes.

**Keep Up**

It is wise to constantly research the marketplace. Read up on the investments you have made, and seek out sources that are up to date with both market trends and world economic conditions.

## 4.2 Reasons Why You Should Invest Your Money

It is not enough to have just a savings account. Though it is crucial to know that saving money is only one aspect of the overall picture. For starters, wise savers open a savings account or invest in a money market account to accumulate enough emergency funds. However, investing in the financial markets has a number of potential benefits after three to six months' worth of accessible savings have been accumulated.

A successful strategy to invest your money and possibly increase your wealth is to open an investment account. Your money may see value growth that exceeds inflation if you make wise investment decisions. The key reasons why investing has a bigger growth potential are the efficiency of compounding and the risk-return trade-off.

Some reasons why you need to invest your money are seen below:

**Increase Your Income**

Even if you do not have $100,000,000 to invest, your money can still benefit from the same opportunities as other people. Make sure that the money you work so hard to earn is working as hard for you as it can.

**Become Independent and Self-Reliant**

You could be better able to live the lifestyle you want once you have amassed wealth. You have the power to transform your life from one of constraints to one of possibilities.

## Give Your Heirs a Legacy

Your wealth can have a significant impact on your heirs by giving them access to educational opportunities, funding for business ventures, or financial assistance for their grandchildren.

## Support Movements You Care About

Having money can be a useful instrument for making a meaningful difference in the world. Therefore, you can utilize your riches to influence positive changes in your neighborhood or around the world, regardless of your love for the environment, the arts, or human welfare.

The choice to invest is an admission that there are dangers involved. Not every investment will succeed, and some may even turn a loss. However, without risk, there would be no chance to perhaps earn the greater returns that can support your wealth-building efforts. Consider keeping your investments broadly diversified to reduce investment risk, taking into account your individual risk appetite, time horizon, and financial goal. Do not forget that diversity is a strategy to assist manage investment risk. If security prices fall, it does not completely eliminate the danger of loss.

A financial expert may be able to help you on your path to wealth creation because investing can be challenging.

## 4.3 Invest in Yourself

The best investment you ever make might be in yourself by learning to invest in yourself. It produces not only future benefits but also, frequently, an immediate pay-off. Putting emphasis on investing in both personal and professional growth is the surest method to improve life quality and be successful, productive, and fulfilled. The amount of work you put into constantly investing in yourself will determine how well your life turns out, both now and in the future.

Putting money into your personal development will increase your self-esteem and confidence in your talents. By concentrating on your personal growth, you will not only acquire new knowledge and abilities but also gain a greater understanding of who you are. You'll gain a better understanding of your particular collection of abilities, principles, and interests, as well as how you might apply them to accomplish your objectives. According to research, people who spend at least five hours each week studying are more likely to feel as though their lives have meaning and purpose, which in turn raises their level of pleasure and wellbeing.

Your career will greatly benefit from making the time investment in yourself, both in the short and long terms. In general, improving your skill set can increase your market value whether you're looking for a promotion in your present position or searching for a new job because you are your biggest asset.

Technical proficiency and academic credentials are still crucial, but the top abilities that companies are seeking nowadays are emotional intelligence, communication skills, and creative thinking.

## Chapter 5: Conscious Spending

## 5.1 Effective Ways to Lower Your Expenses

**Know where your money is going**

Financial confidence is being demonstrated to increase when you write down your weekly expenses. You should therefore keep track of your spending to improve your financial stability. Budgeting is useful in this situation. Your monthly spending and your monthly revenue are the first two crucial figures in every budget. Make a budget that records both your income and your expenses. Having a better understanding of how your money is coming in and going out will help you examine your spending and saving habits over time to spot any patterns.

**Make spending categories**

Make a list of your necessities, wants, and personal values the things that are most important to you after setting a budget. Having a clear understanding of your values can assist you in creating a budget for the things that matter most to you, such as starting a business, contributing to the community, or spending time with your loved ones. Depending on your budget, necessities like rent or mortgage payments, groceries, and utilities will probably take precedence over "fun" expenditures. Look at your spending and ask yourself what you can live without for the time being.

**Spend Money on Important Things**

Even though "fun" purchases might be put on the back burner, you should still include them in your budget because they can still be beneficial to your mental health. The things that are in line with your personal values can still be saved for and purchased, such as trips to surprise family members or tuition for a course you wish to take.

You'll feel better and more content with your purchase decisions if you stick to a budget based on your beliefs, and saving money won't feel like a chore.

## Take Advantage of "Monthlies"

Here are some suggestions to reduce monthly recurring costs since they might mount up. Try working out at home instead of the gym, stopping your subscription to the public transportation system, or choosing groceries over takeout or delivery, which can be more expensive. Make a list and determine what you are willing to postpone.

## Cut Off Impulsive Purchases

The majority of Americans (50%) admit to making unnecessary purchases. If your social media scrolling frequently results in "Thanks for shopping," look for techniques to reduce impulsive purchases. Your financial goals may be derailed by impulsive purchases. Be aware of the things that set this off the most for you, such as emails with 50% off sales from your favorite apparel retailer. Create an email filter to hide them from you unless you are actively shopping. When an urge does occur, go back to your list of principles. Click "Remove from cart" and keep your attention on the wider picture if the purchase doesn't fit with your objectives.

## Where You Can, Avoid Paying Interest

If you have a mortgage or vehicle loan, you are aware of the importance of interest in your monthly payment. If money is tight, see if refinancing can help you eventually pay less in interest. Alternately, if your income hasn't changed, you might want to think about delaying other spending in order to make one or more additional principal payments. This shortens the loan's term, increases equity, and lowers interest costs.

## Think About Postponing

Some financial organizations have provided forbearance options including delaying auto loans at this time. Consult your bank if necessary to learn what possibilities they can offer you. This will minimize your immediate expenses even though you will have to pay eventually. However, if at all possible, avoid deferment because you will continue to accrue interest during that time, which will make your final payment higher.

## Chapter 6: Multiple Stream of Incomes

You most likely have one principal source of income if you are like most people. While there is nothing wrong with that, it might be dangerous to rely solely on one source of income. What would occur, for instance, if you lost your work or your main source of money stopped? This was a common experience throughout the epidemic, along with job losses and furloughs.

For generations, large corporations have diversified their sources of revenue. For the purpose of creating additional revenue streams, they diversify their business operations. Any business may diversify. There are more methods of generating money outside diversification, known as the "seven streams of revenue";

**Income from Work**

Your main source of revenue comes from your job, which you earn. We all start here, and many of us never move on. The term "just over broke" was used to describe how severely restricting earned money is for most people. In other words, your income is barely enough to get by. It's true that some jobs offer extraordinarily high income, but they are the exceptions, not the rule. Take chances and transition into profit income if you want to leave your job and launch your own business.

**Profit Income**

You generate income based on profit by charging more for a good or service than it costs. You may start a retail business and sell goods, provide expert services and charge for your time, or do both.

The transition from earned income to profit income is one of the most difficult tasks, although many employees strive for it. There are risks and challenges involved in going independent or starting your own business.

**Interest Earnings**

You or your company is losing money if you have extra money in your bank account. You can invest your money in a variety of ways to generate passive income. Consider investing it in a savings plan and utilizing compound interest to generate a passive income. Another secure investment that yields money is purchasing government bonds.

**Gains from Dividends**

Purchasing shares in a firm entitles you to dividend payments and makes you a part owner of the business. Business investments made at the right time can produce excellent passive income sources.

## Rental Income

Investing in real estate is a great method to safeguard your funds and earn rental income. This income source has two drawbacks. Unless it is a component of an investment strategy, it first requires a sizable upfront investment. Second, if you might need the money urgently, this is not for you because releasing the money might be expensive and time-consuming.

## Investment Income

You may earn money from buying and selling assets, which is known as capital gains. For instance, the capital gain is $20 if you purchase stocks and shares for $100 and later sell them for $120. Since each nation has distinct regulations, it is imperative to first speak with an accountant concerning capital gains. Your profit can be completely eliminated by capital gains tax, depending on the sole asset.

**Royalty Income**

Designing, creating, or manufacturing something special and charging individuals and companies to use it results in this passive income stream. The best example is musicians. A specific label, like Virgin Records, is usually where musicians sign their contracts. The musician recording, record production, marketing, and sales are all paid for by the record label. Every time an album is sold or broadcast to the general public, the musicians get paid a royalty. The royalties from performing Elton John's music generate millions of dollars for well-known musicians.

It's crucial to have several sources of income for this reason. This manner, you'll have backup streams in case one of them runs dry.

# Chapter 7: Keep Record of Your Progress

## 7.1 Effective Ways to Keep Your Records

Garbage in, garbage out is a saying that initially pertained to computer hardware and software, but it also applies to the financial aspect of your organization. If you want to be successful, you must have a financial structure that eliminates waste.

In order for you to understand how you're doing and know when to act, your system should provide you with accurate and valuable information.

**Take Note of the Data**

Nothing exists if it is not present. As you begin your business, develop the practice of taking notes on everything. It will eventually start to happen on its own.

The process's most crucial stage is "capture," which is also the most challenging step. Keep a record of all the money you spend on your business and all the money you get from sales. Even if you usually pay your business back when you pay personal spending out of that account, be careful to keep personal and business expenses apart.

**Verify that the Data Is Accurate and Up to Date.**

Every two weeks or so, spend an hour going through everything you've recorded to make sure it's all ready for recording. Establish a definite time for yourself to meet at the conclusion of each alternate week to "check everything," such as every other Friday.

Waiting too long will make it harder to remember the knowledge.

**Save the Information by Recording It.**

Your financial data must be recorded in a format that may be used. You can either have your bookkeeper record everything you've checked, or you can record it all yourself. Make a habit of it. Enter the data in a spreadsheet or accounting program. With online software, you and your bookkeeper may both view the data and interact with it as needed. Ensure that everything is recorded each month so that you may evaluate it.

**Utilize what you already know**

Establish trigger points at which the knowledge compels action. Liabilities likely indicate rising assets or expenses if they increase each month for three months in a row on your balance sheet. Reduce your spending. If you notice on your income statement that a particular expense is rising as a percentage of sales, consider why. If the rise is required, you might want to reduce spending on other costs to keep your profit level the same.

# Chapter 8: Common Habits of Successful People

## Organization

Organization is one of the traits of successful people that is commonly emphasized. Planning is a part of this structure, along with establishing priorities and objectives.

## Relaxation

It is interesting to see that avoiding distractions and relaxing through meditation are two other successful people's practices that are frequently cited. Those that are more structured naturally relax more readily, thus for some people, relaxation may come more naturally than on purpose.

## Taking Initiative

The inescapable "action" habit comes in third on the list of successful people's habits. Prioritization, organization, and planning are crucial, but without action, a plan is nothing more than a potential solution.

People who are successful take action swiftly and frequently. James Clear also claims that despite it seeming contradictory, people act (start, at any rate) before they feel prepared. Successful people take the crucial initial step even if it seems absurd, while others find excuses not to.

## Personal Care

The second habit of successful people is personal care, including food, exercise, and hygiene. Some people's personal care entails an intricate routine and a highly regimented way of life.

## Networking

Successful people understand the need of networking for idea exchange. They are also aware of the importance of cooperation and teamwork, both of which are probably present when you network. Successful people understand the value of associating with other successful people.

## Frugality

Being frugal does not mean being stingy. The tendency of practicing thrift with resources and money is known as frugality. Additionally, it is a tendency of being frugal. Avoiding waste is the first step in learning to be frugal, which naturally leads to efficiency. Successful people avoid spending too much money. Instead, they bargain and shop around. Through the straightforward action of saving more money than they spend, they achieve financial success.

**Sharing**

Successful people have a practice of giving, whether through monetary donations to charities or the exchange of ideas. They understand the importance of giving to others, and the majority of them think that success should lead to more than just their own personal wealth accumulation.

**Reading**

The fact that successful people read should not be overlooked. Even though they also read for enjoyment, the majority of people read for information or understanding.

# Conclusion

Even though it is difficult, it is possible to become a billionaire before the age of 30, and this is especially true in this era of booming entrepreneurship. One must first begin directing the majority of their income toward the purchase of assets if they want to become wealthy. The hardest part is right here. Spending tracking and budgeting are the methods to use.

The first several days will be very difficult. But keep in mind that psychology is everything. Simply develop the mentality to live on less money. Additionally, it's important to monitor your rising net worth. What keeps one motivated is this.